GEOMETRIC DESIGNS
CD-ROM and Book

DOVER PUBLICATIONS, INC.
Mineola, New York

The CD-ROM on the inside back cover contains all of the images shown in the book. There is no installation necessary. Just insert the CD into your computer and call the images into your favorite software (refer to the documentation with your software for further instructions). Each image has been scanned at 600 DPI and saved in six different formats—BMP, EPS, GIF, JPEG, PICT, and TIFF. The JPEG and GIF files—the most popular graphics file types used on the Web—are Internet ready.

The "Images" folder on the CD contains a number of different folders. All of the TIFF Images have been placed in one folder, as have all of the PICT, all of the EPS, etc. The images in each of these folders are identical except for file format. Every image has a unique file name in the following format: xxx.xxx. The first 3 or 4 characters of the file name, before the period, correspond to the number printed with the image in the book. The last 3 characters of the file name, after the period, refer to the file format. So, 001.TIF would be the first file in the TIFF folder.

For technical support, contact:
Telephone: 1 (617) 249-0245
Fax: 1 (617) 249-0245
Internet: dover@artimaging.com
World-Wide Web:
http://www.artimaging.com/dover.html

The fastest way to receive technical support is via email or the World-Wide Web.

Bibliographical Note

Geometric Designs CD-ROM and Book is a new selection of images from the following Dover books: *Geometrical Design Coloring Book* by Spyros Horemis, 1973; *Visual Illusions Coloring Book* by Spyros Horemis, 1973; *The 3-D Alphabet Coloring Book* by Jean Larcher, 1978; *Prismatic Design Coloring Book* by Peter Von Thenen, 1978; and *Kaleidoscopic Design Coloring Book* by Lester Kubistal, 1999.

Dover Electronic Clip Art ®

International Standard Book Number: 0-486-99538-0

Manufactured in the United States of America
Dover Publications, Inc., 31 East 2nd Street, Mineola, N.Y. 11501

The images in this CD-ROM and book offer a variety of abstract geometric designs. In some, mere lines on the flat surface of the sheet conjure up a strong conviction of three-dimensionality—even if the solid object suggested by the linear interplay could not possibly exist. In others, multiples of letters of the alphabet interlock, interlace, fuse and merge in unusual patterns. Still others are kaleidoscopic in nature, with delicate and intricate patterns radiating out from a central point. All are visually exciting and appealing and can be used in many different ways.

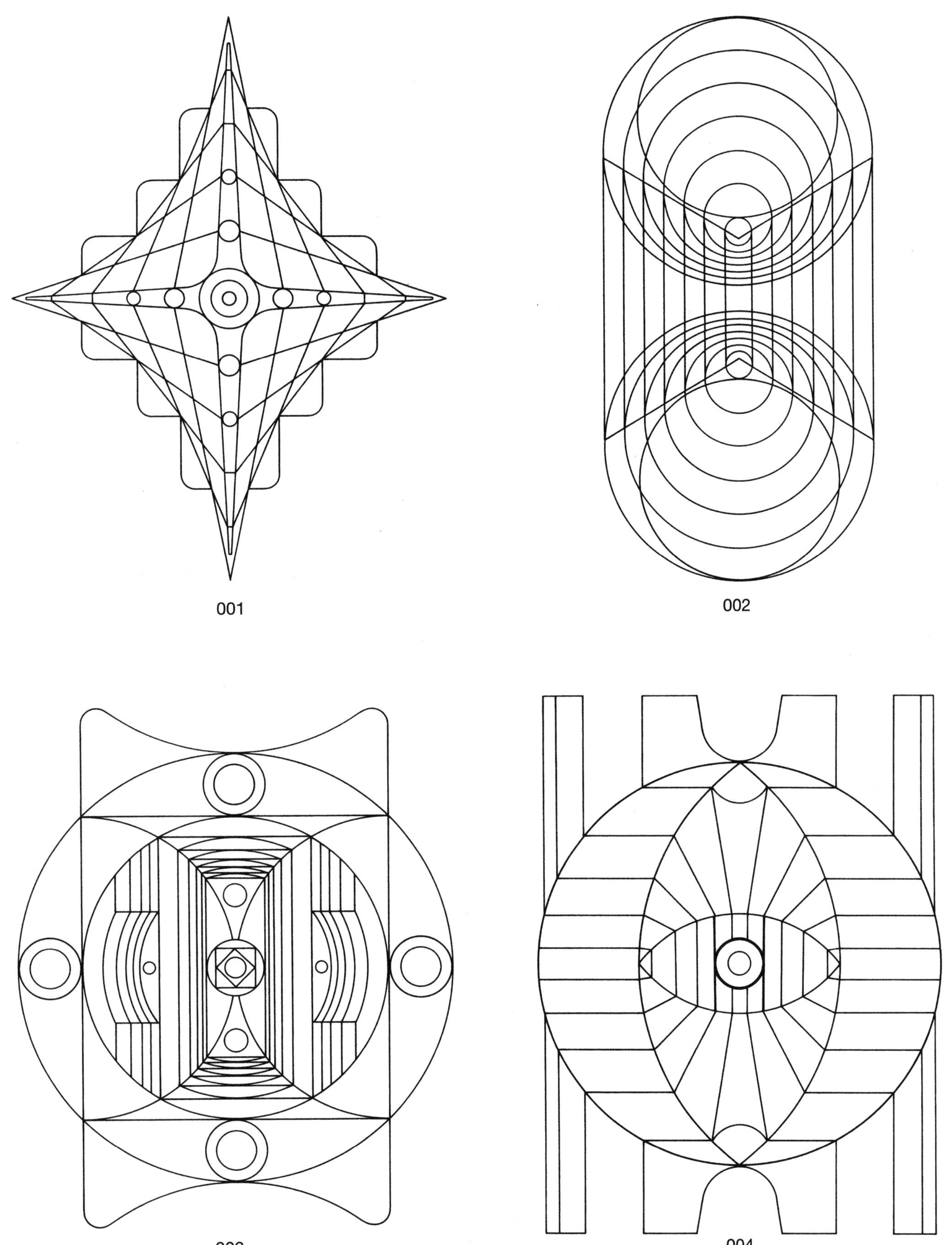

001
002
003
004

005

006

007

008

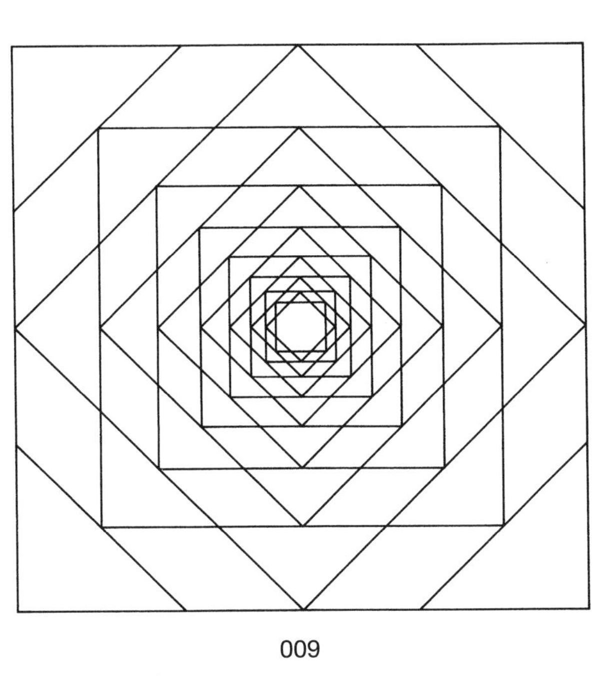

009

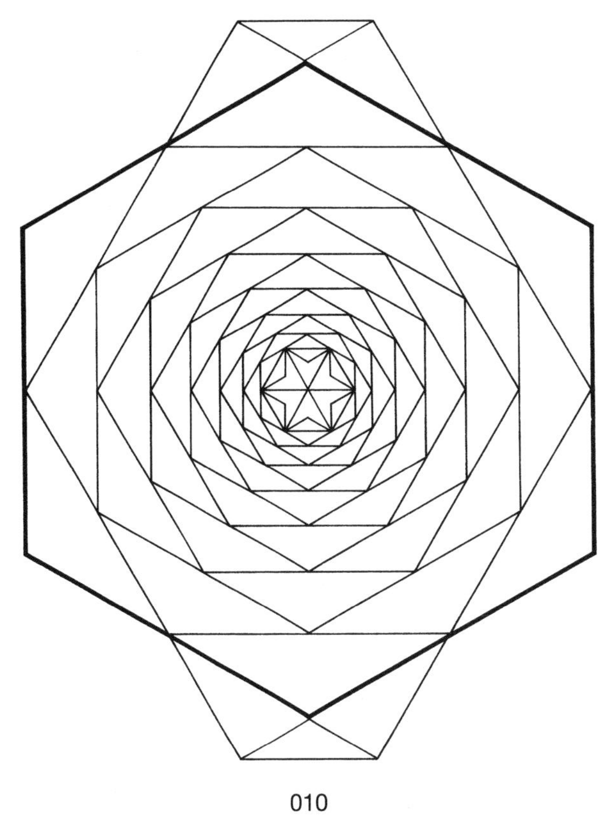

010

011

012

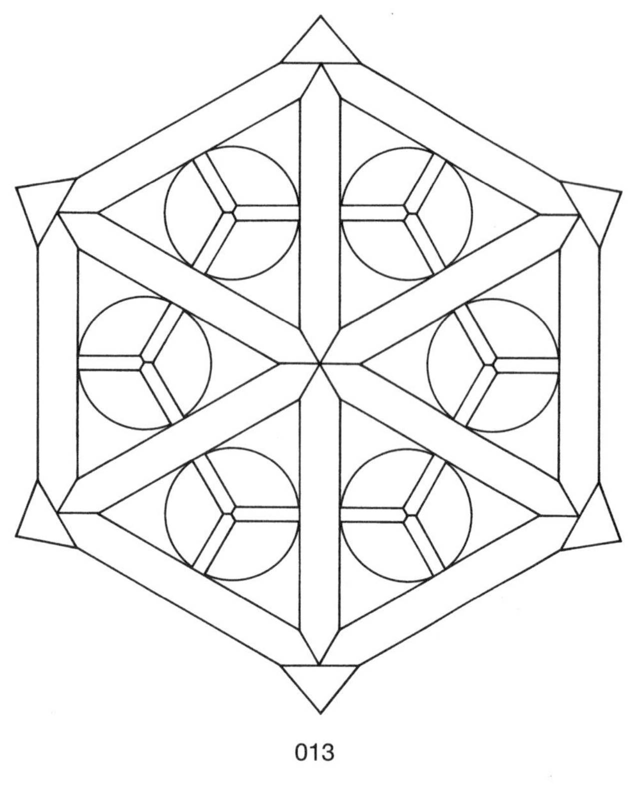

013

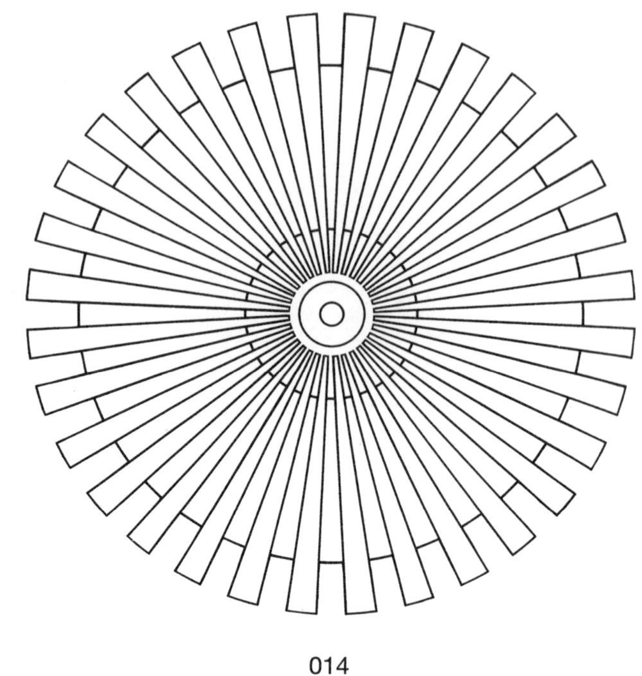

014

015

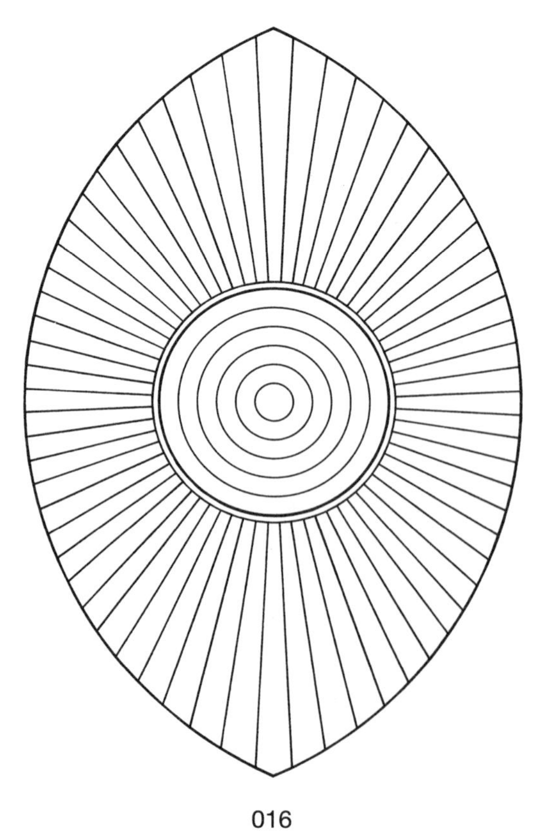

016

017

018

019

020

021

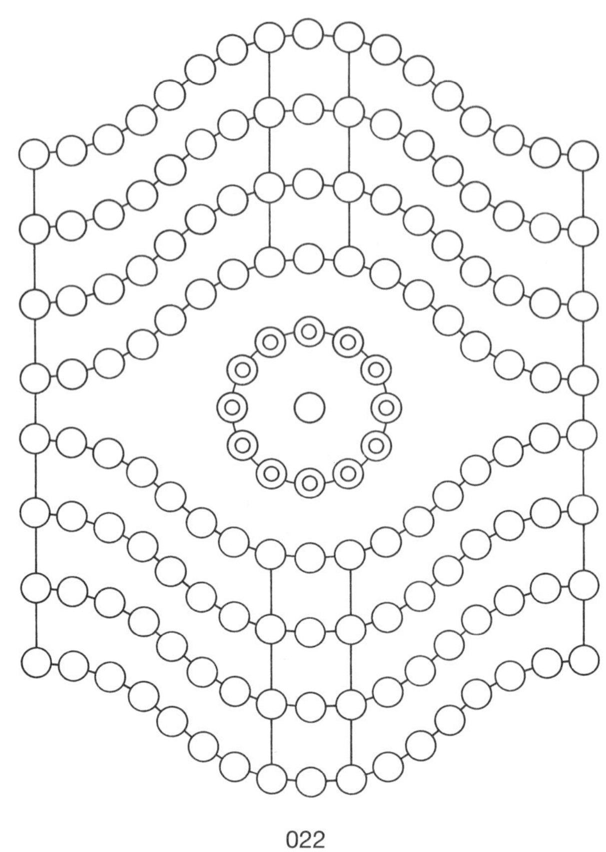

022

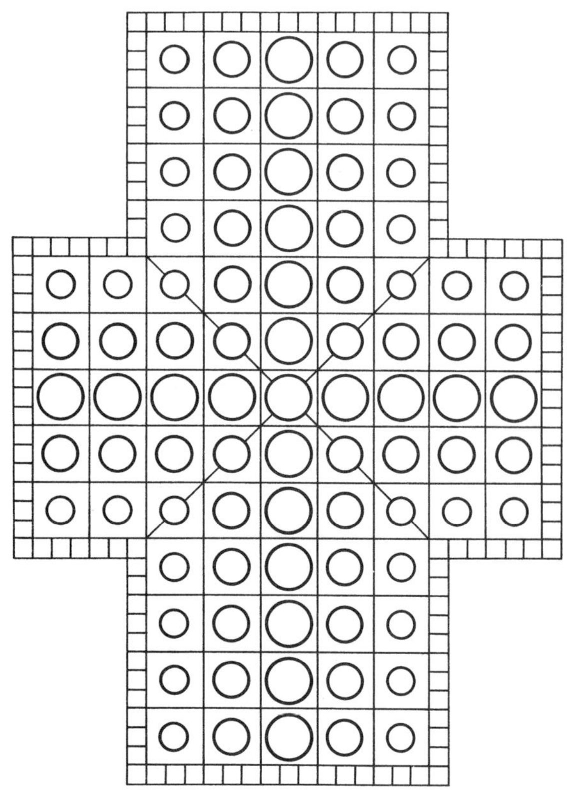

023

024

025

026

027

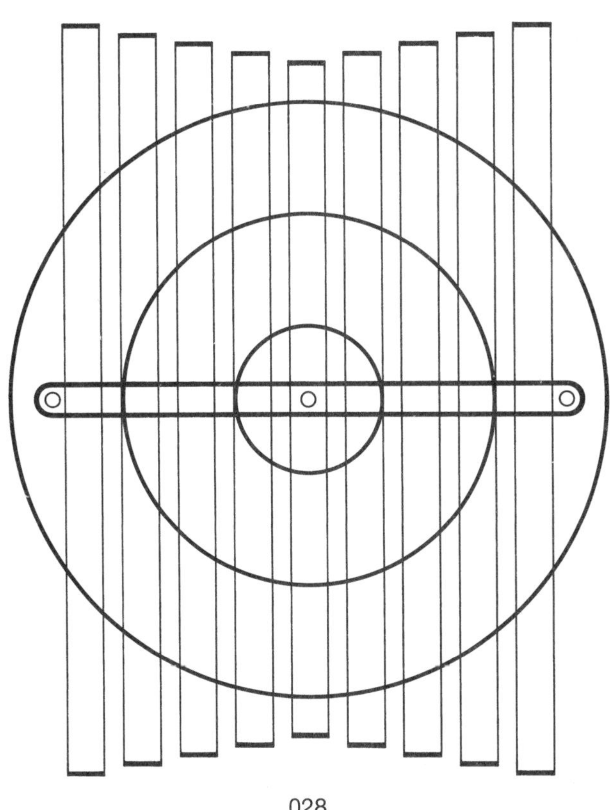

028

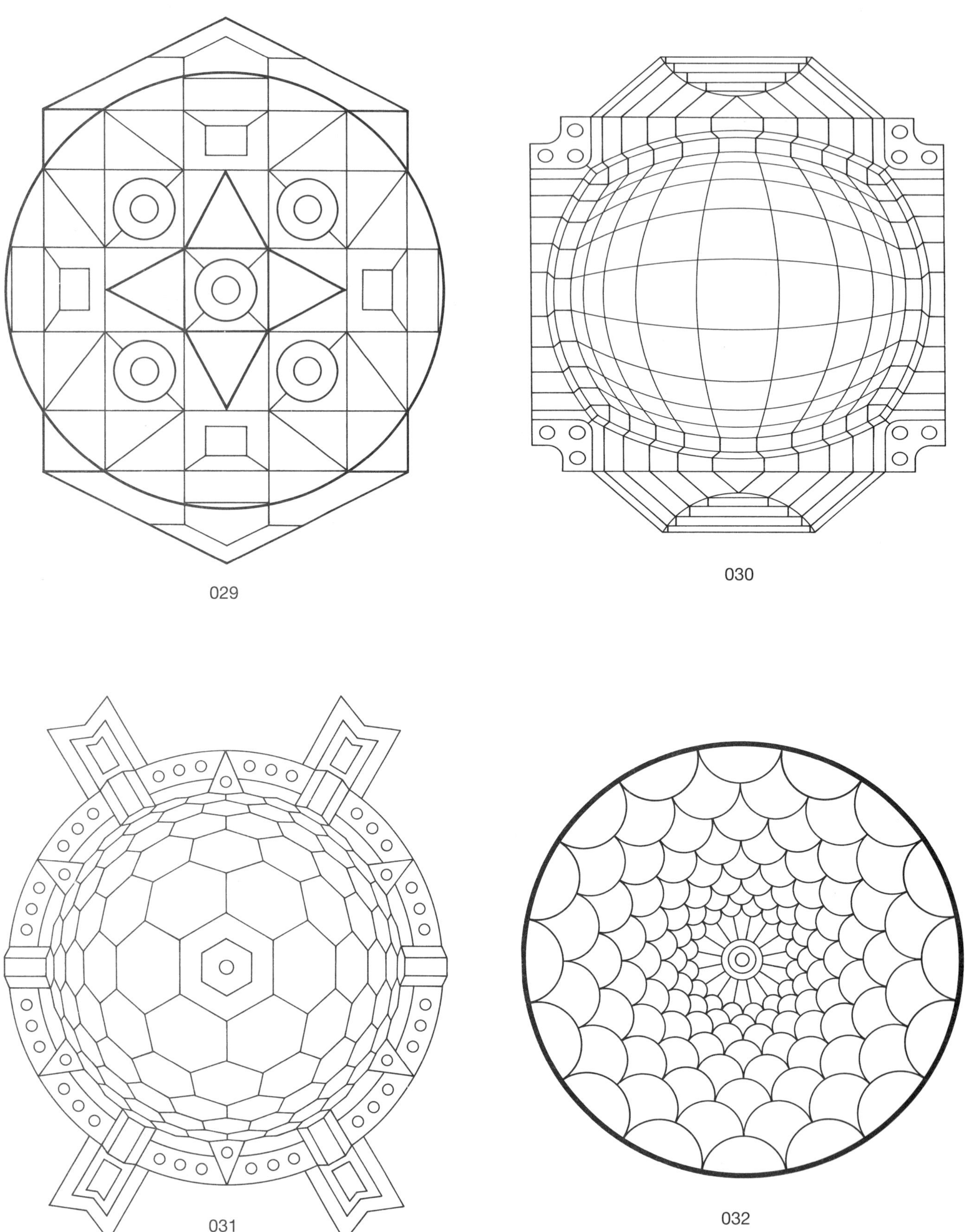

029

030

031

032

033

034

035

036

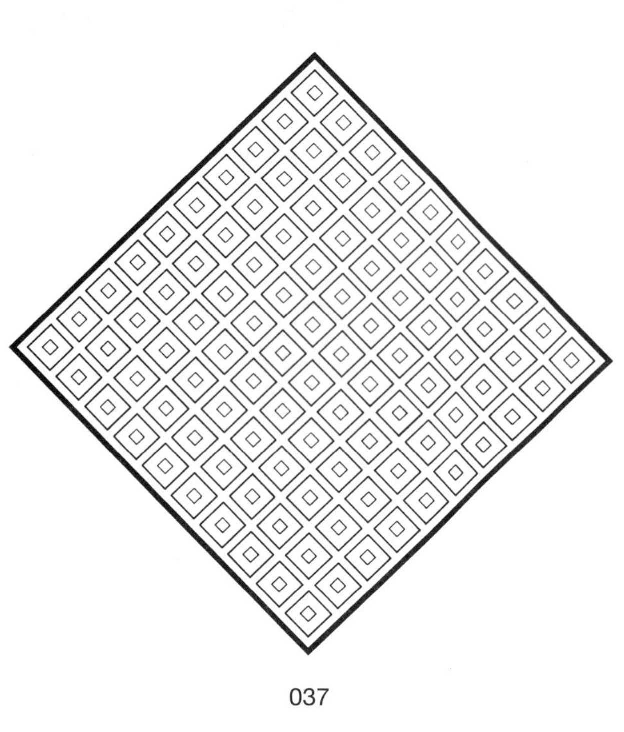

037

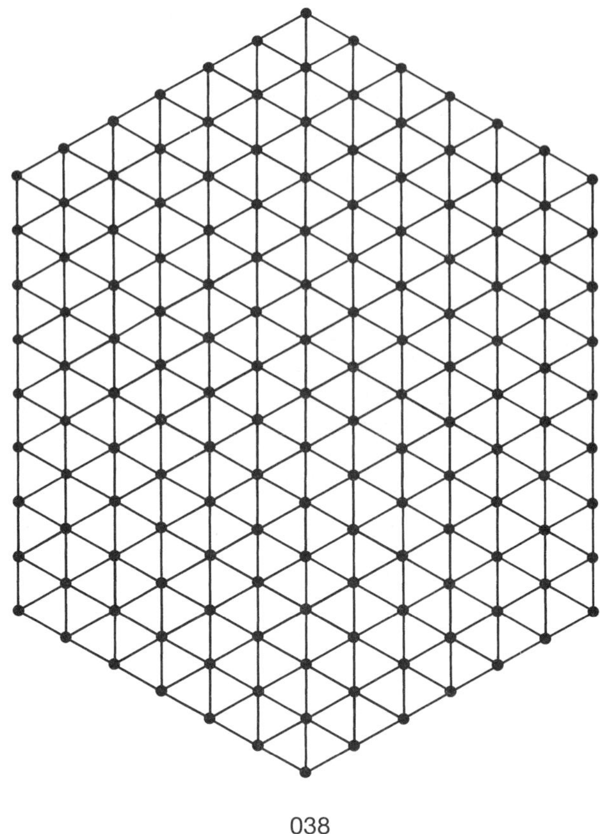

038

039

040

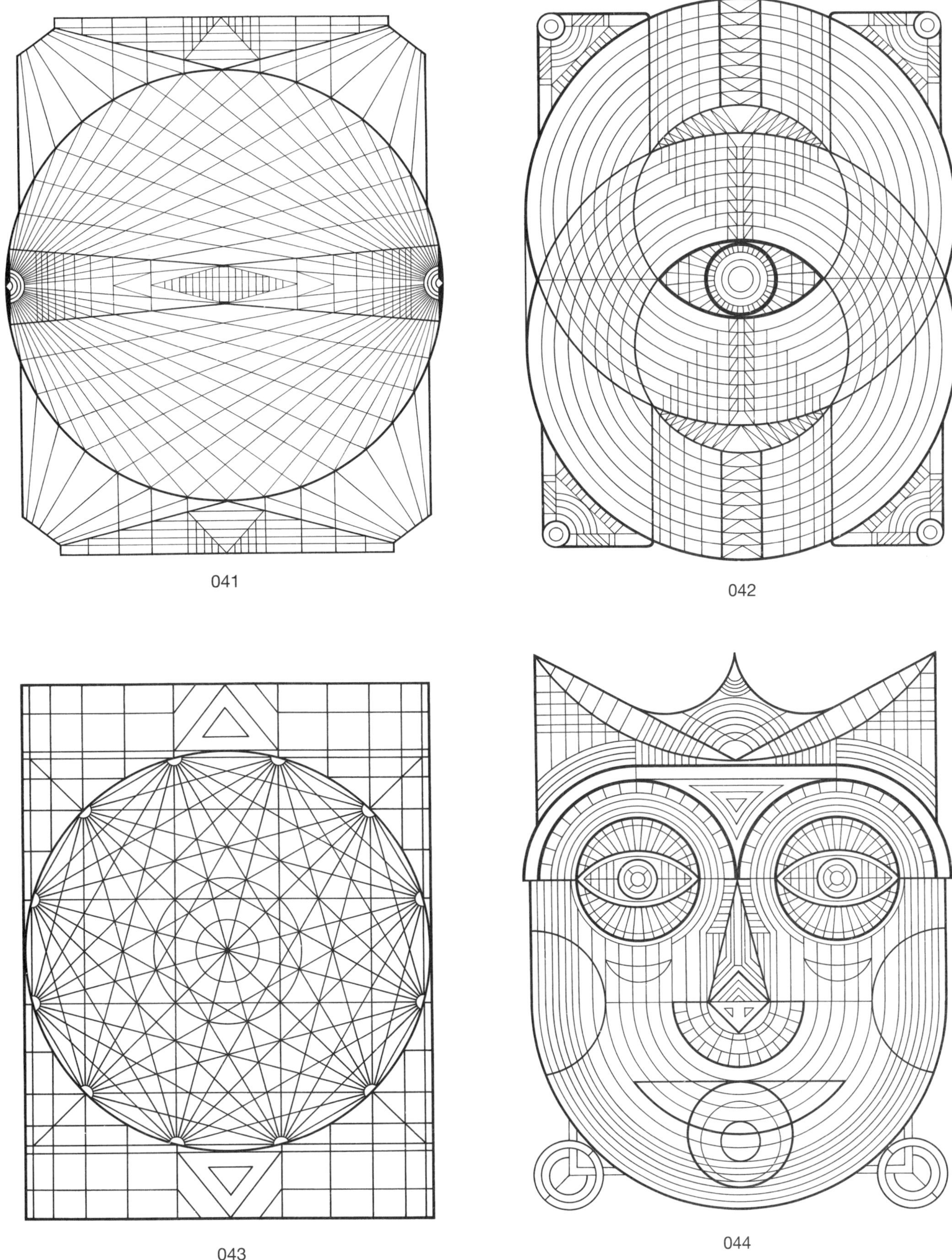

041
042
043
044

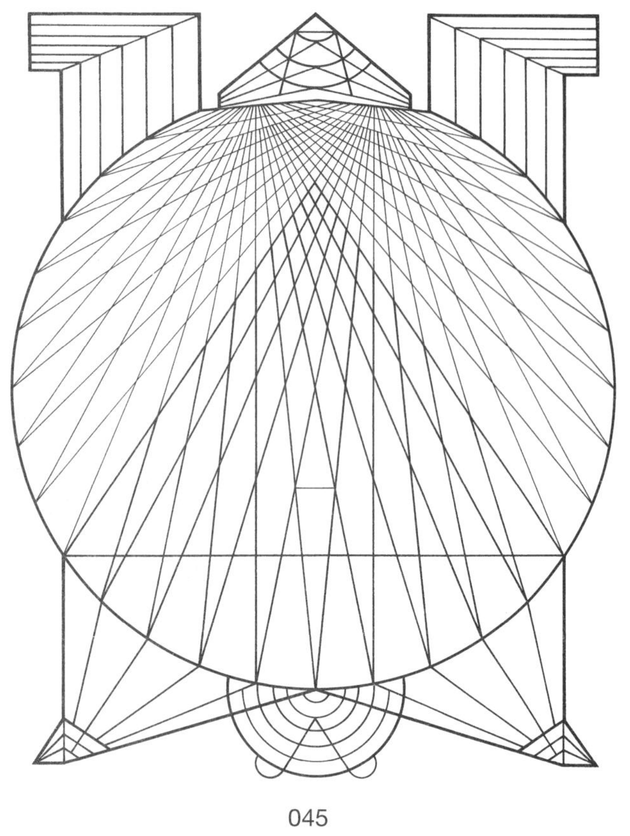

045

046

047

048

049

050

051

052

053

054

055

056

057

058

059

060

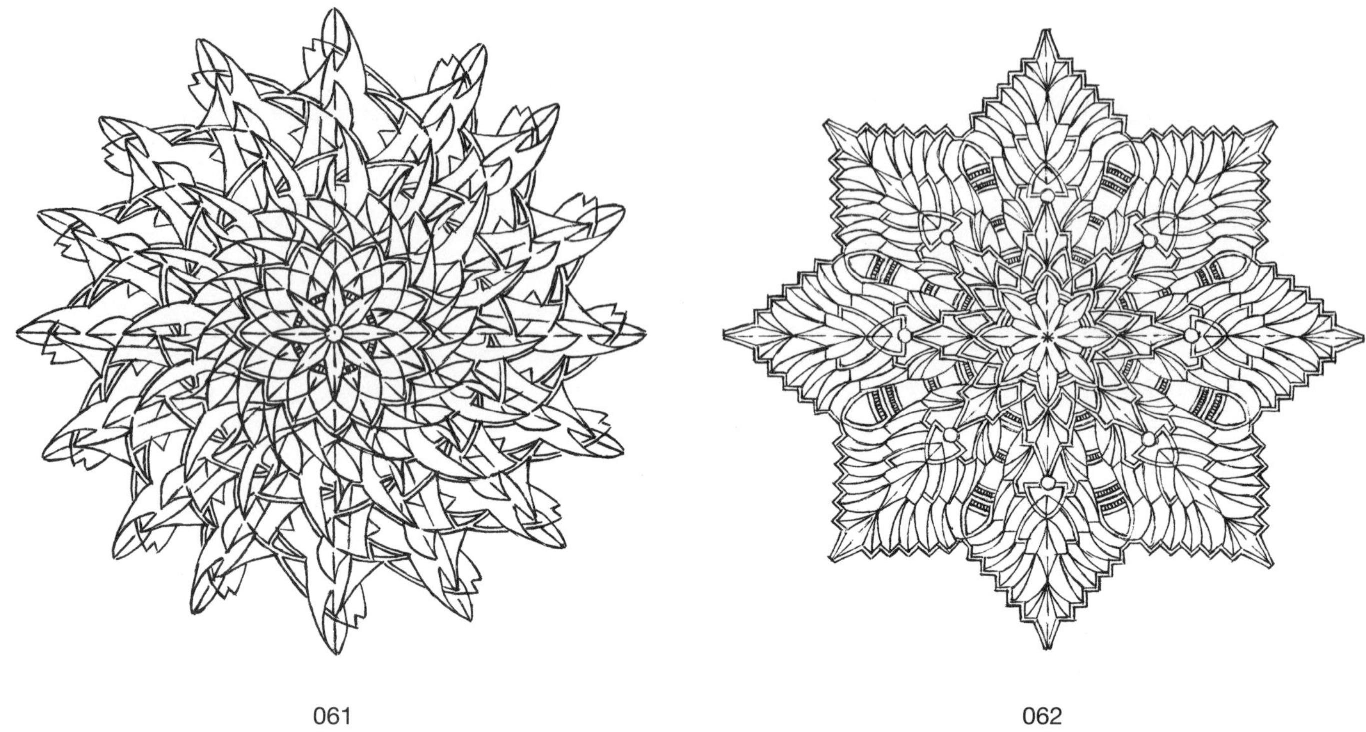

061

062

063

064

065

066

067

068

069

070

071

072

073

074

075

076

077

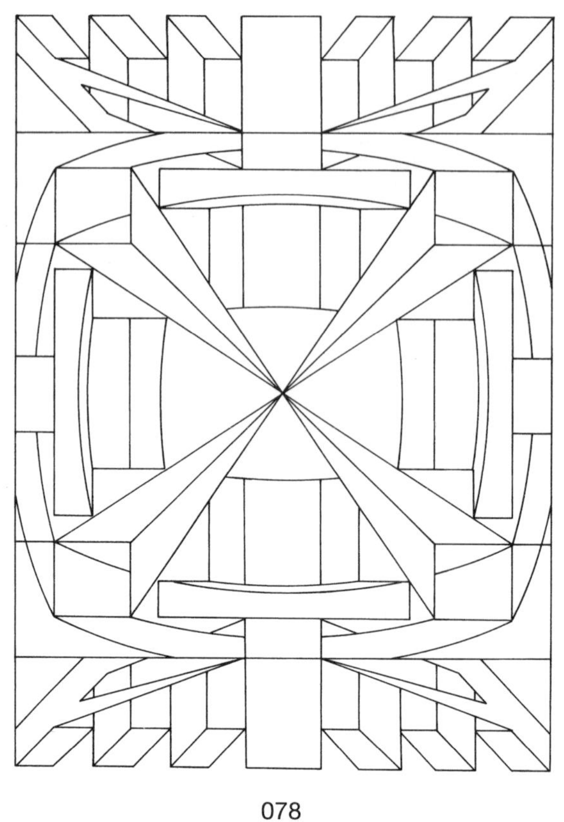

078

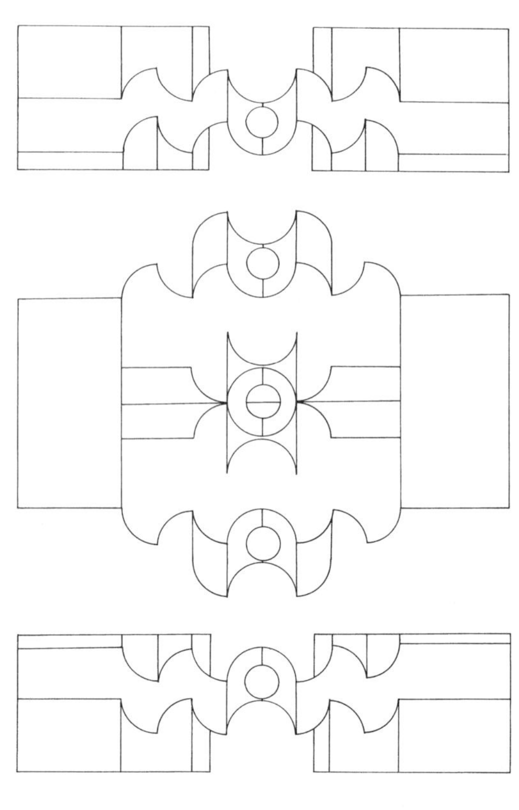

079

080

081

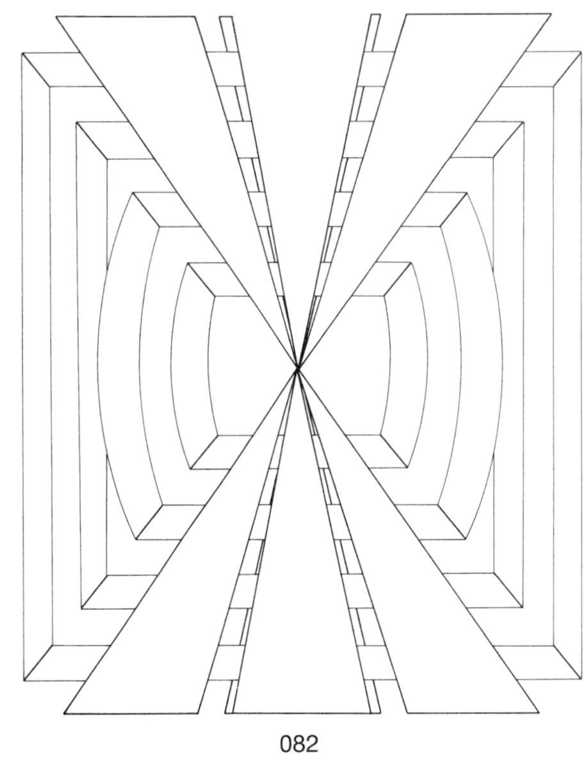

082

083

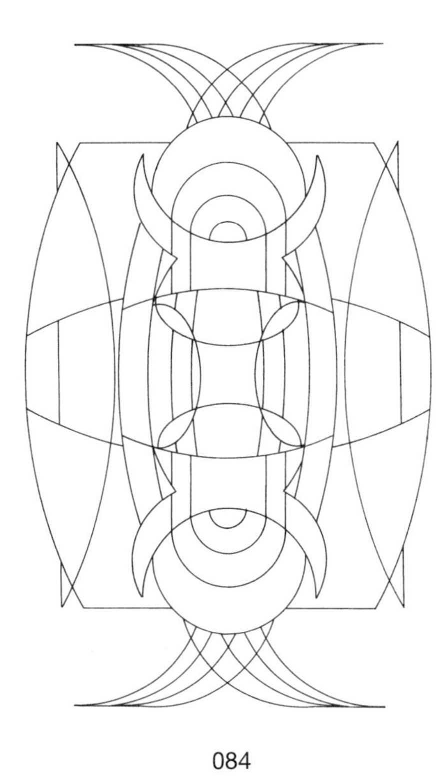

084

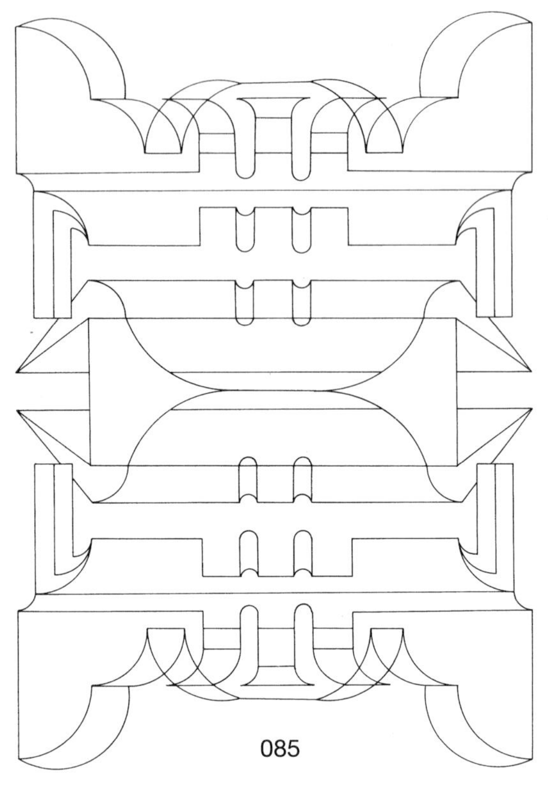

085

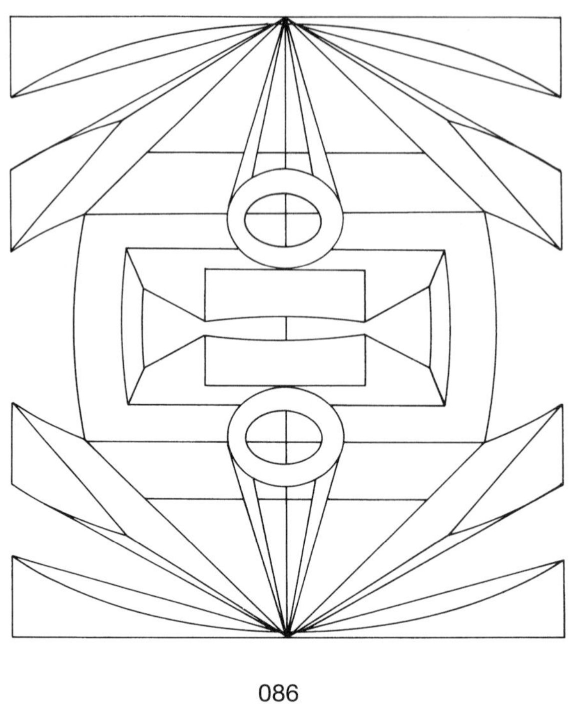

086

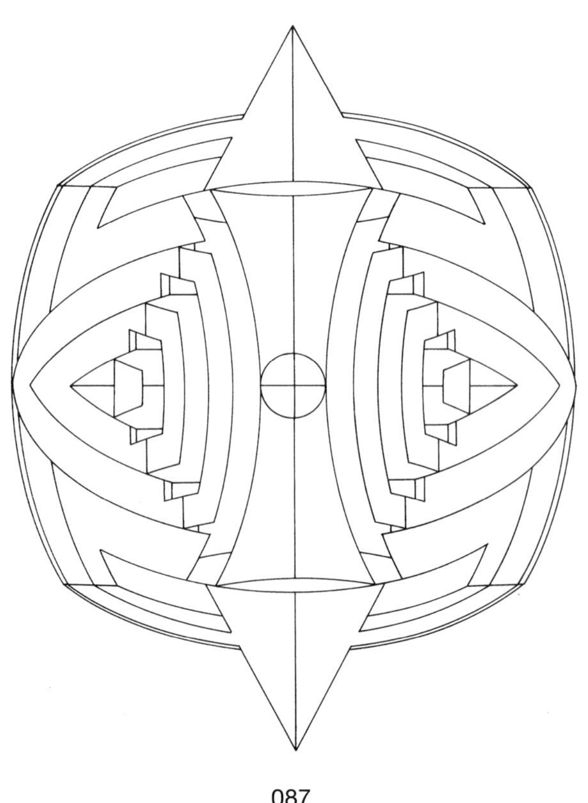

087

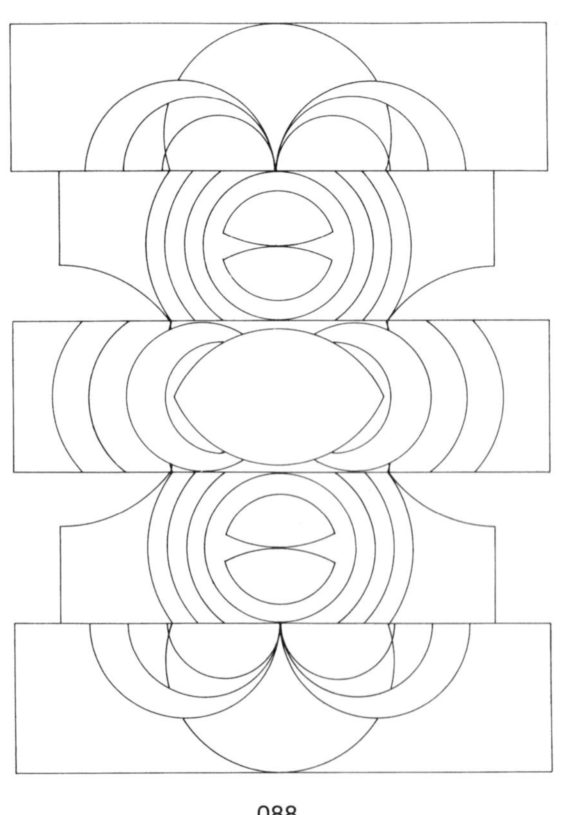

088

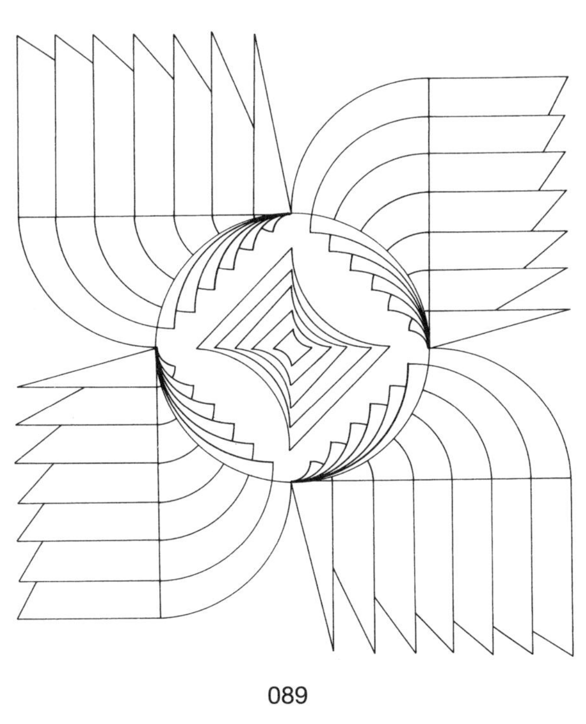

089

090

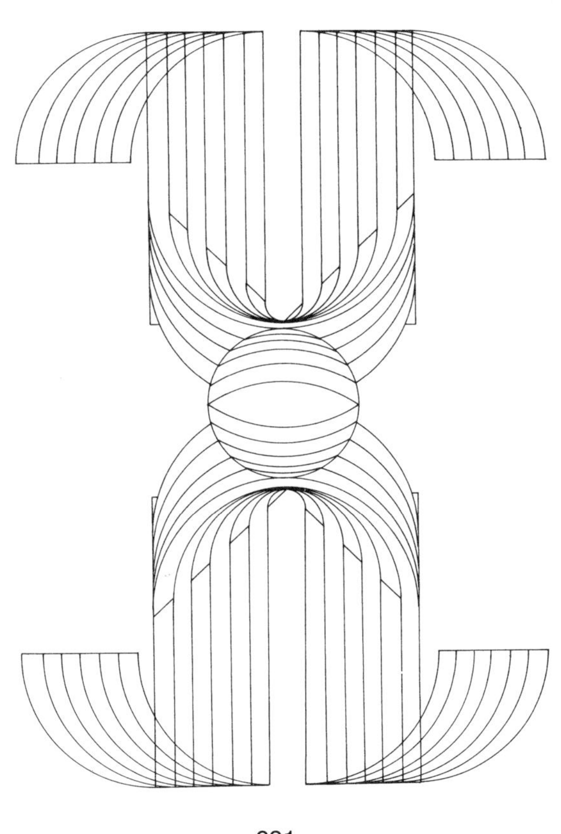

091

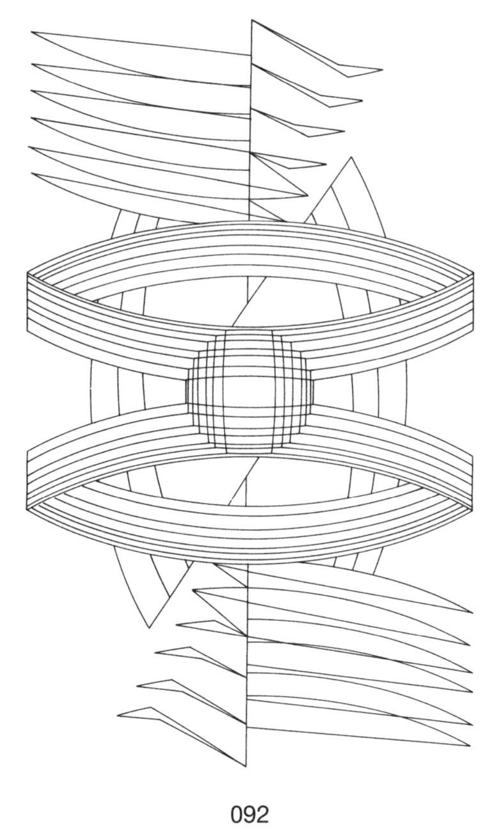

092

093

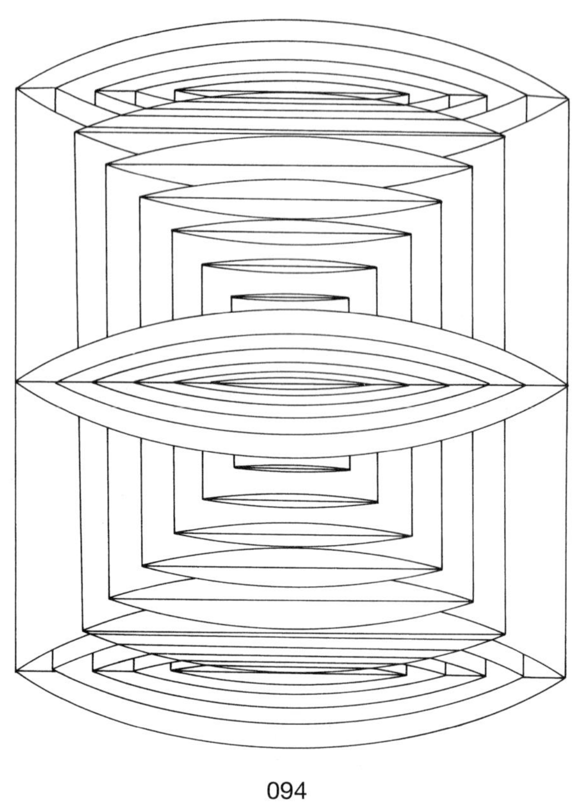

094

095

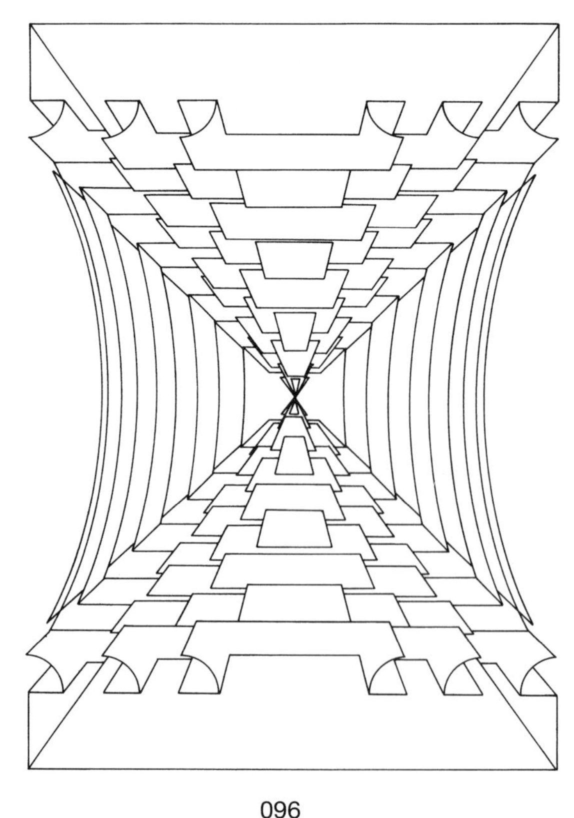

096

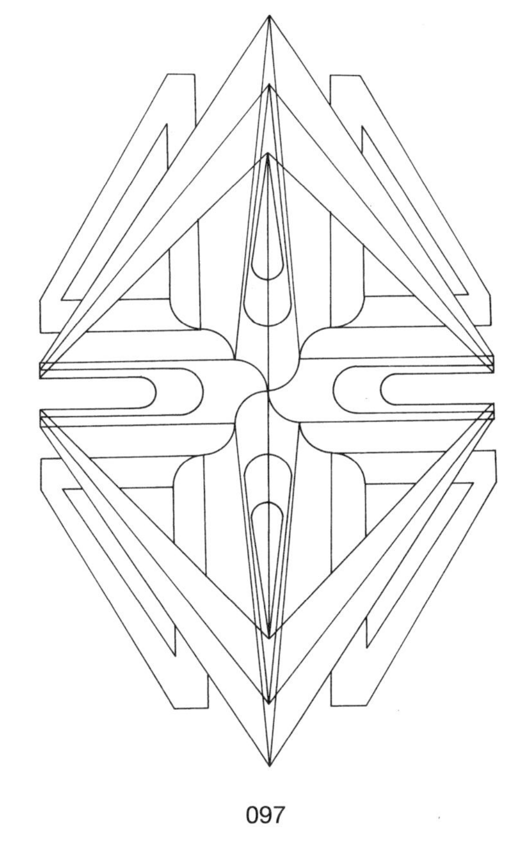

097

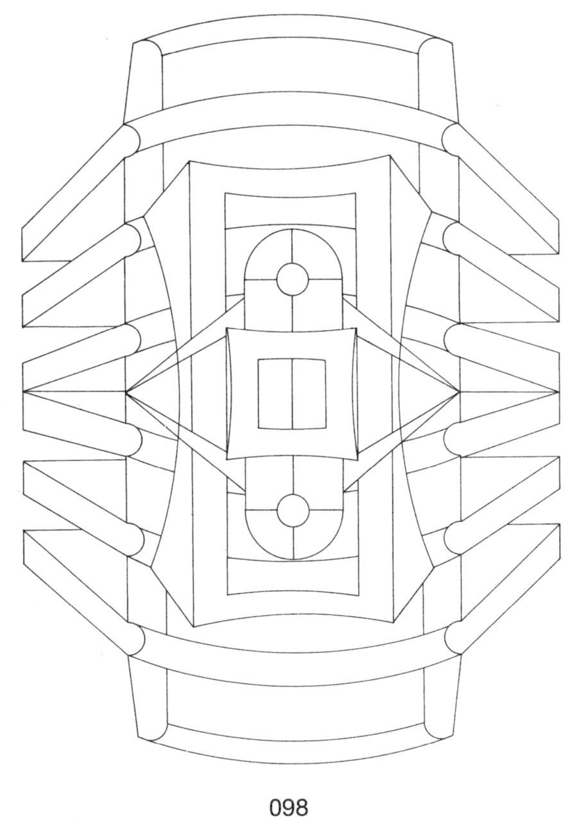

098

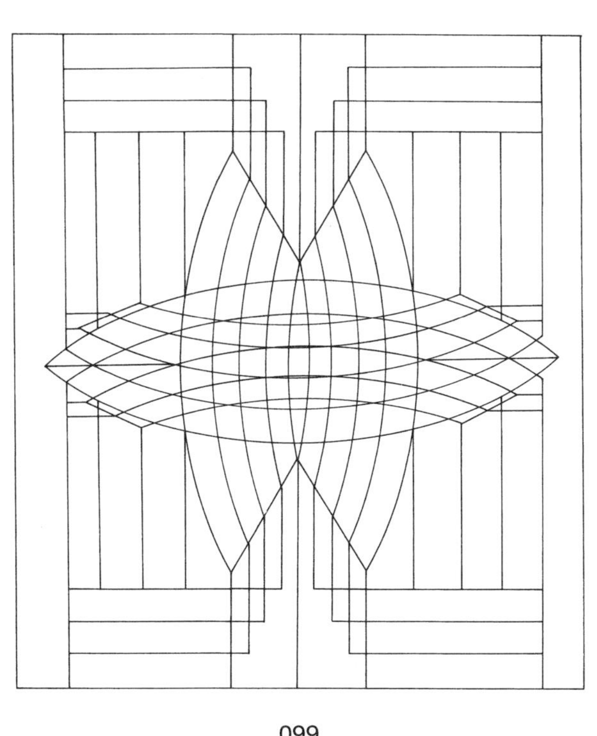

099

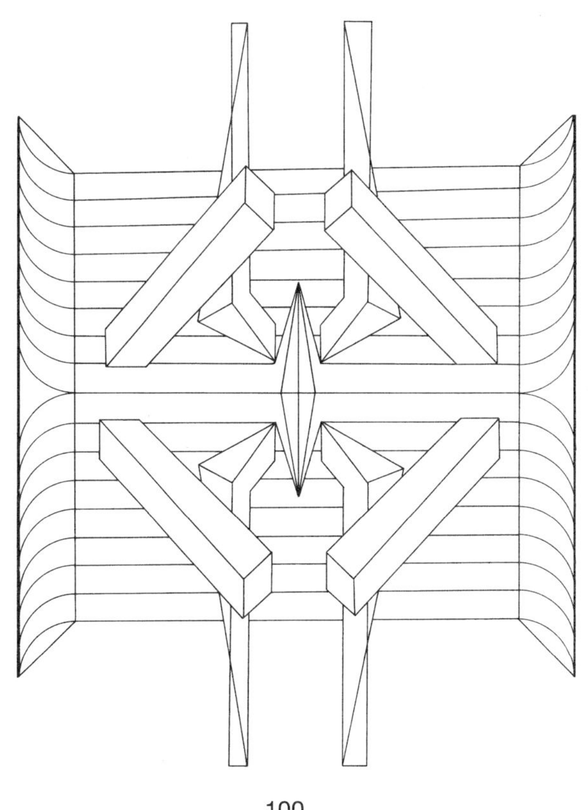

100

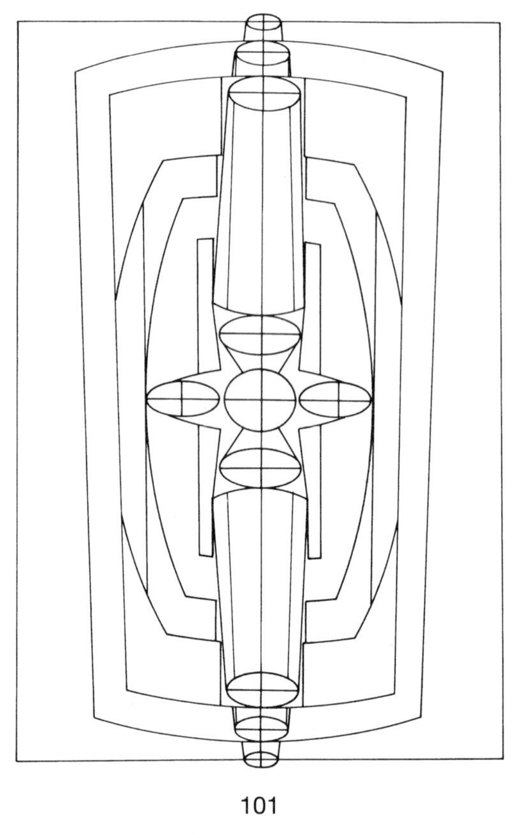

101

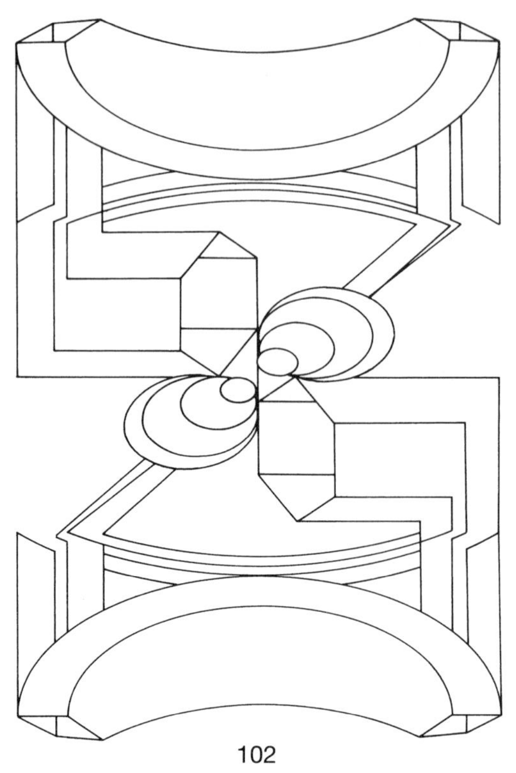

102

103

104

105

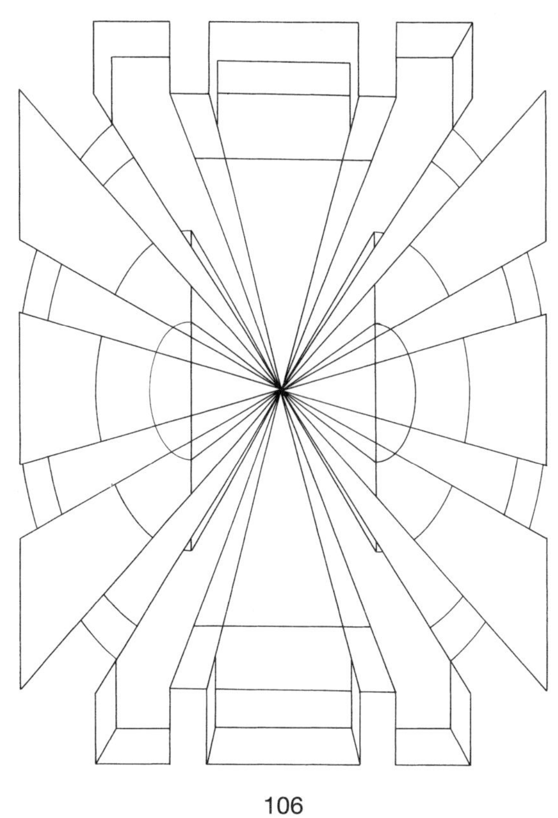

106

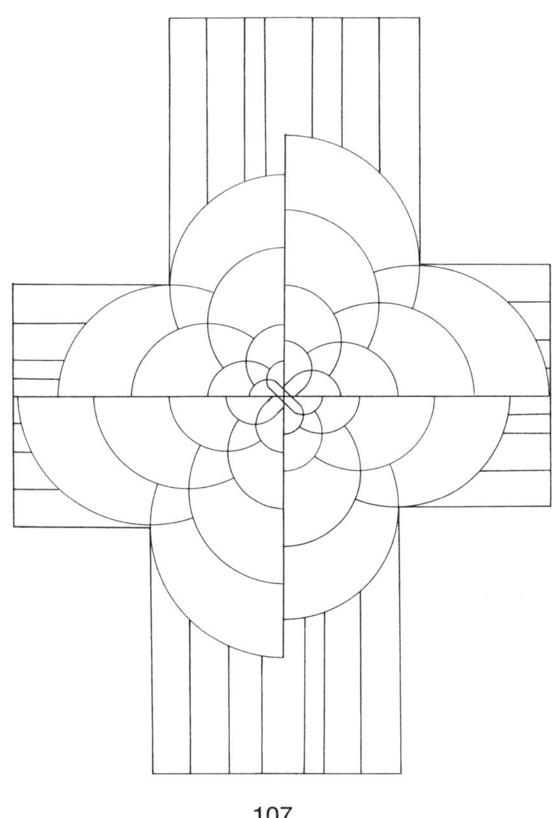

107

108

109

110

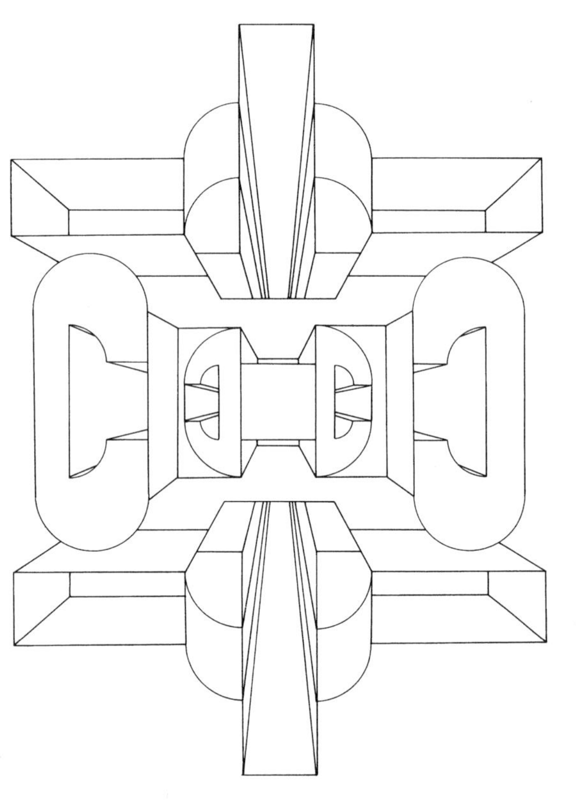

111

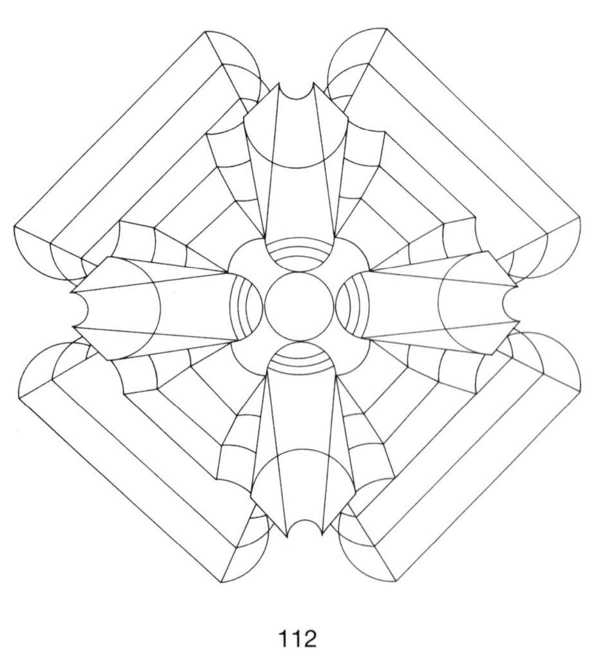

112

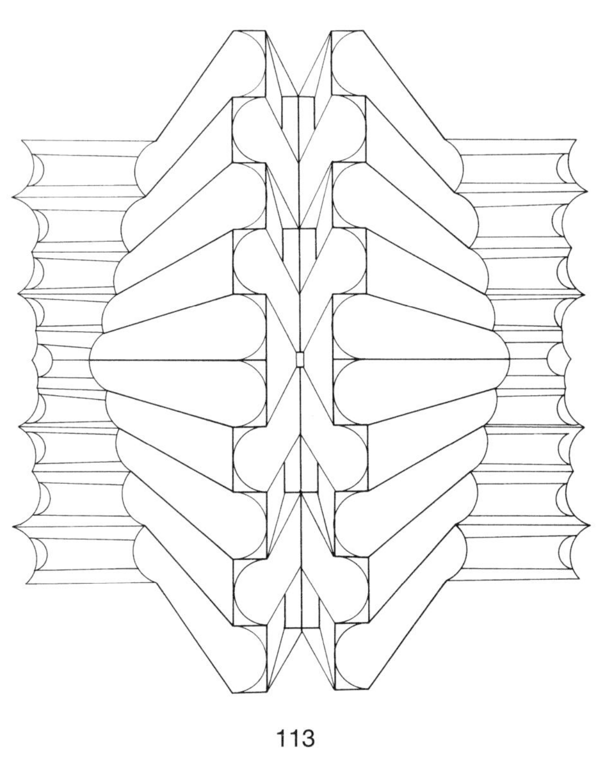

113

114

115

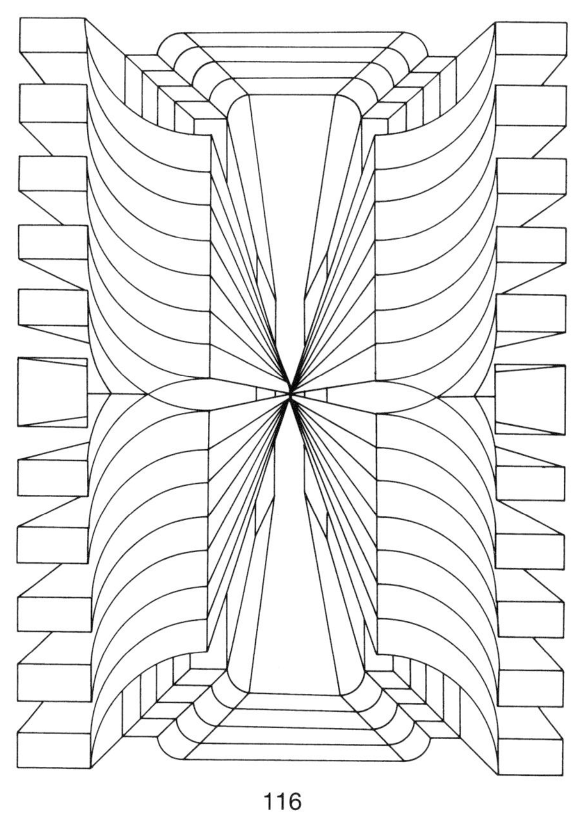

116

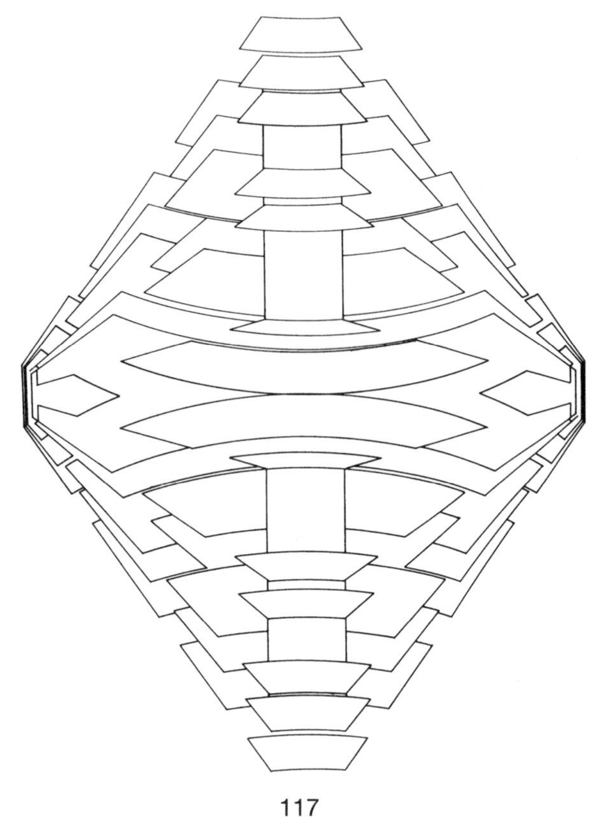

117

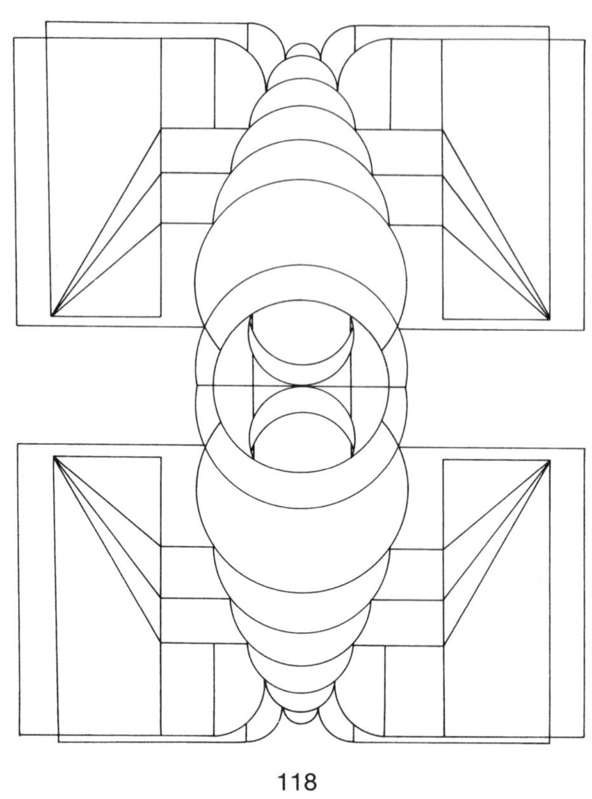

118

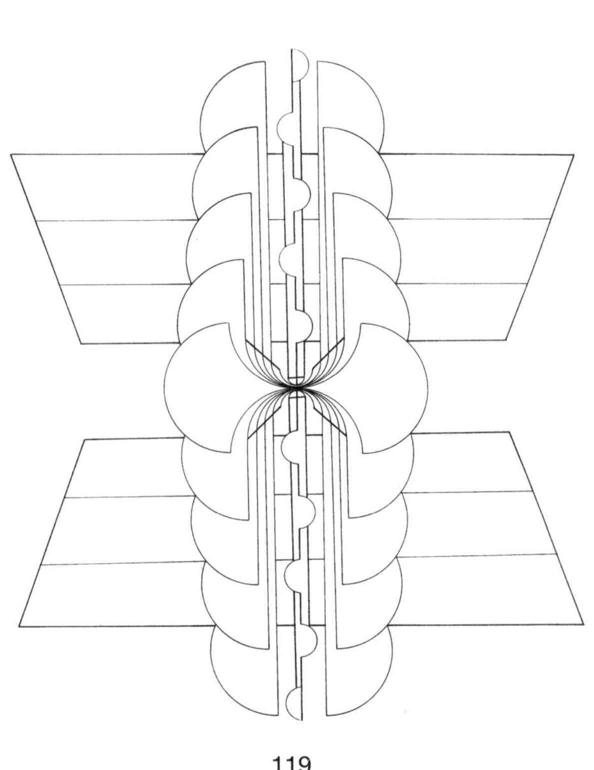

119

120

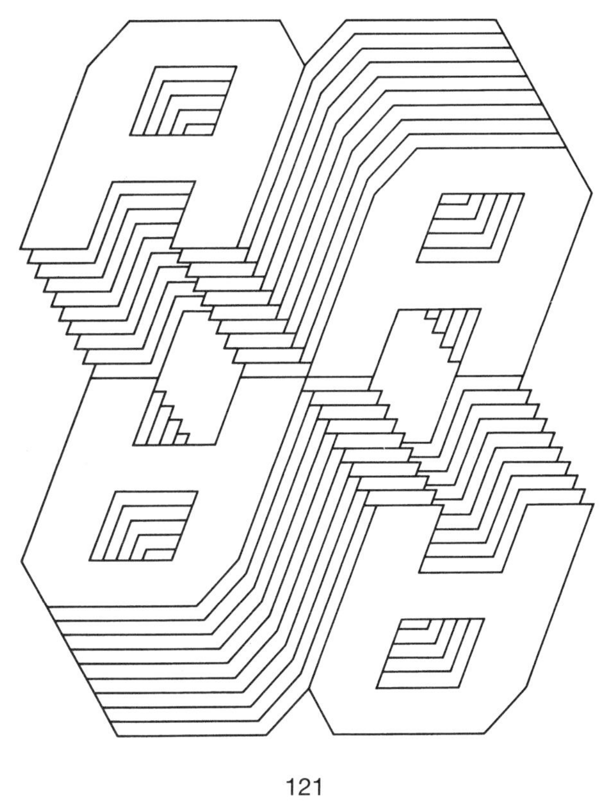

121

122

123

124

125

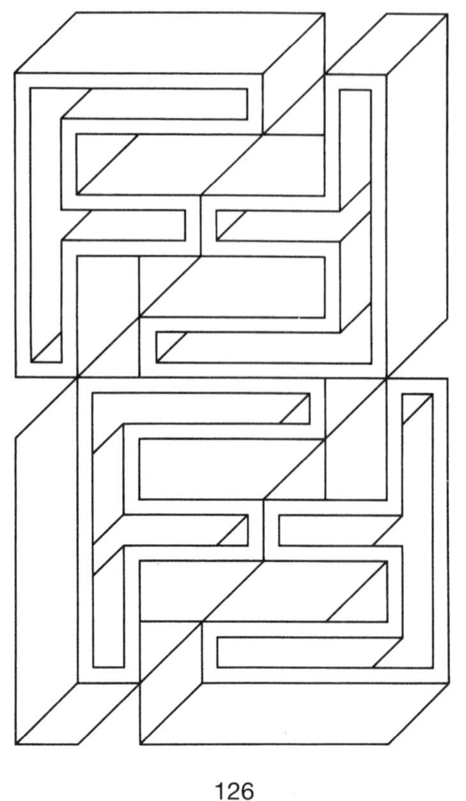

126

127

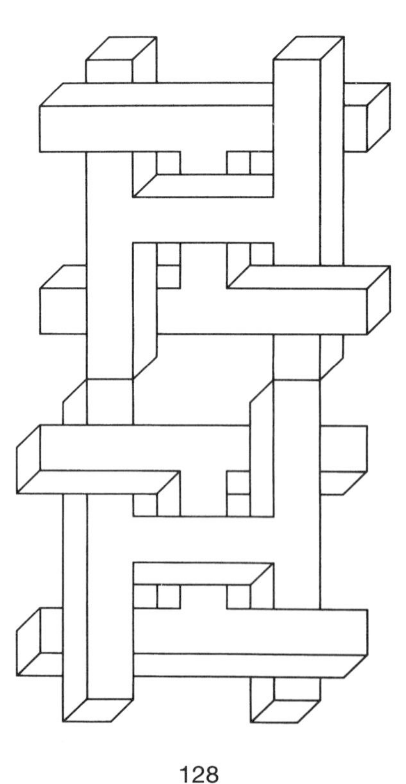

128

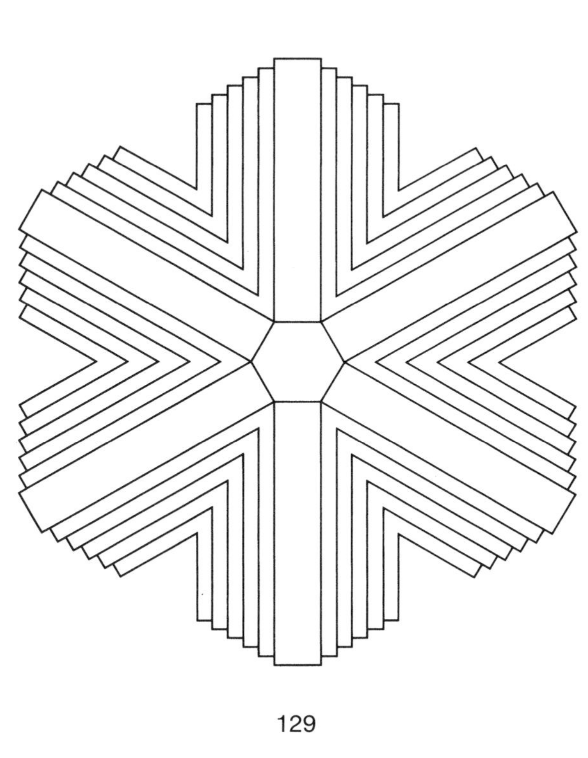

129

130

131

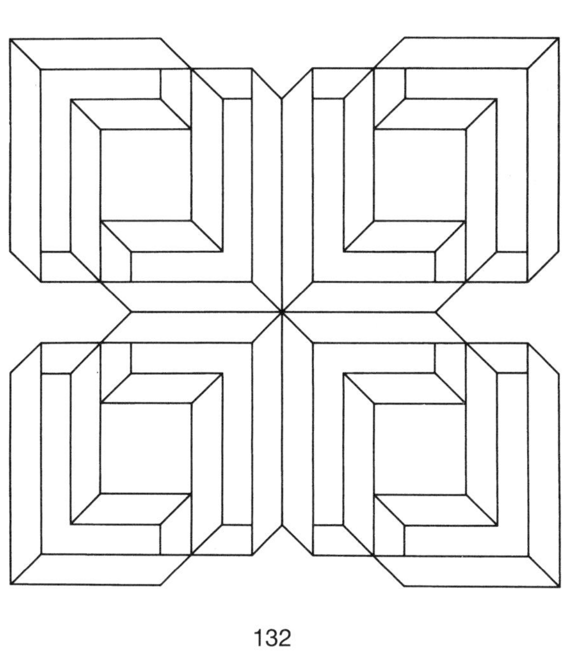

132

133

134

135

136

137

138

139

140

141

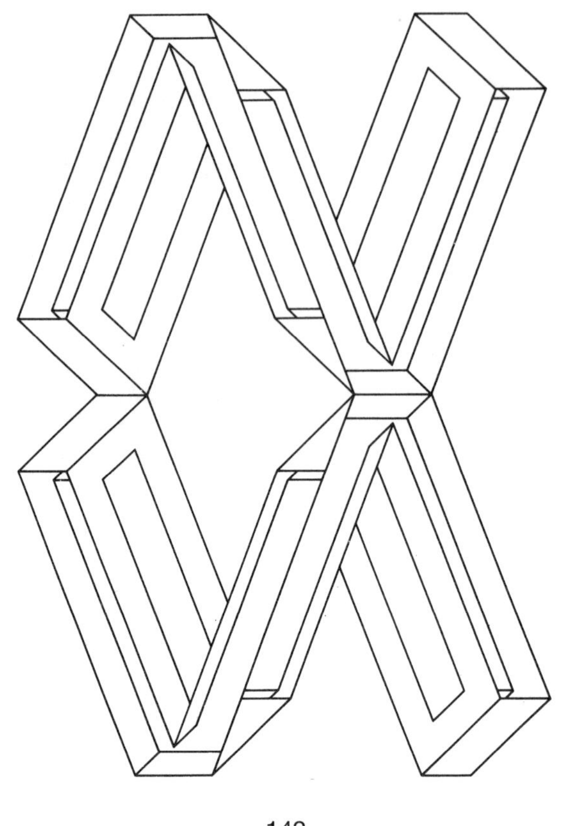

142

143

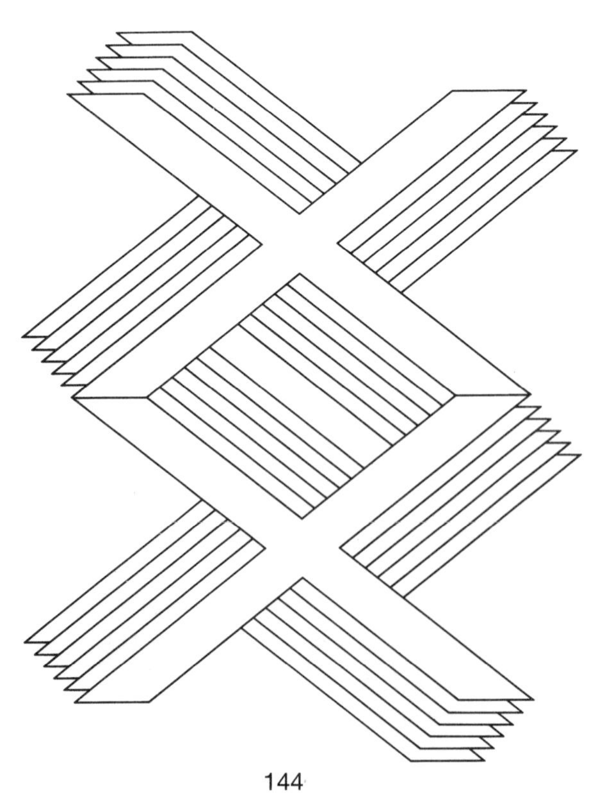

144

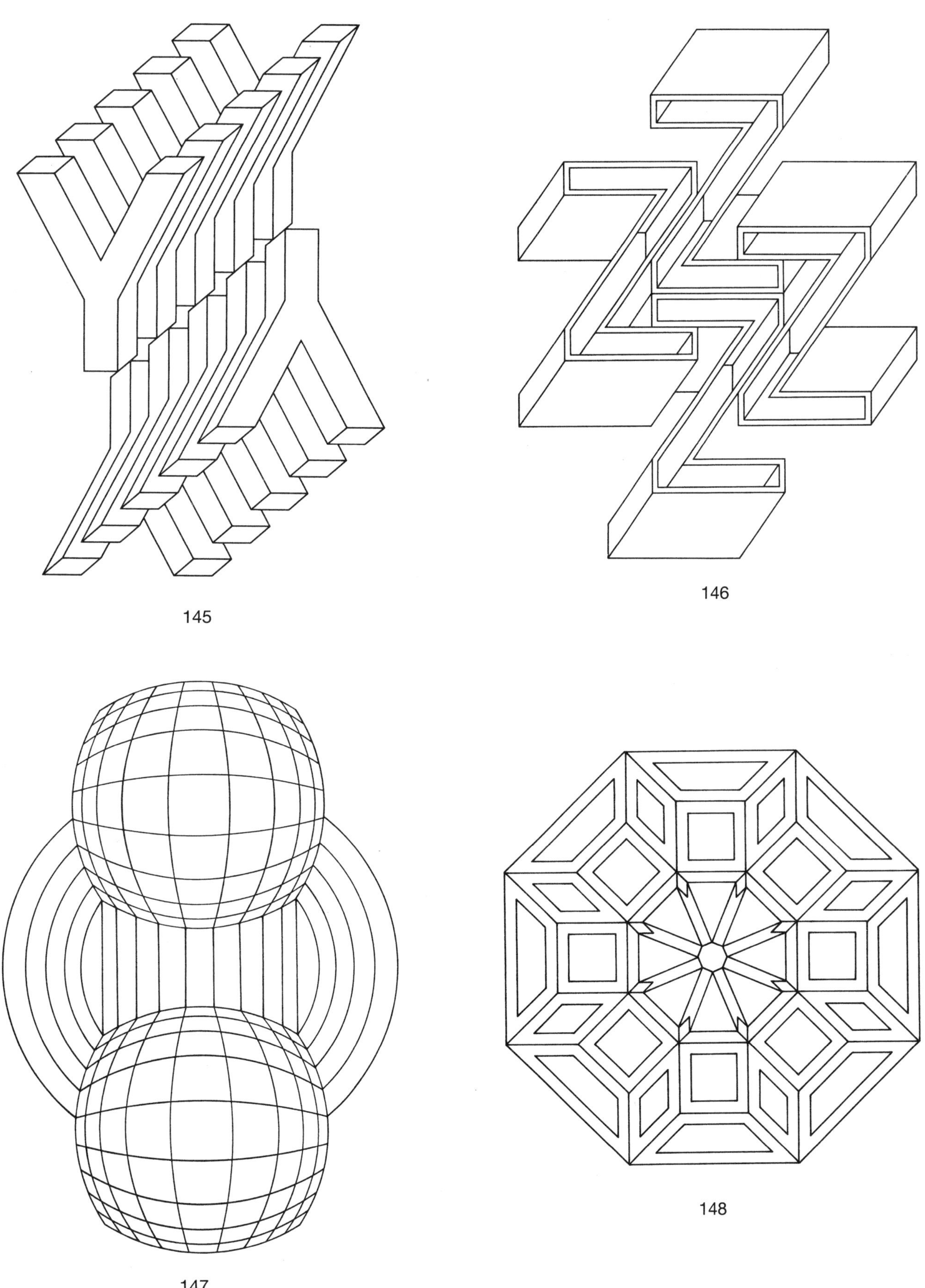

145

146

147

148

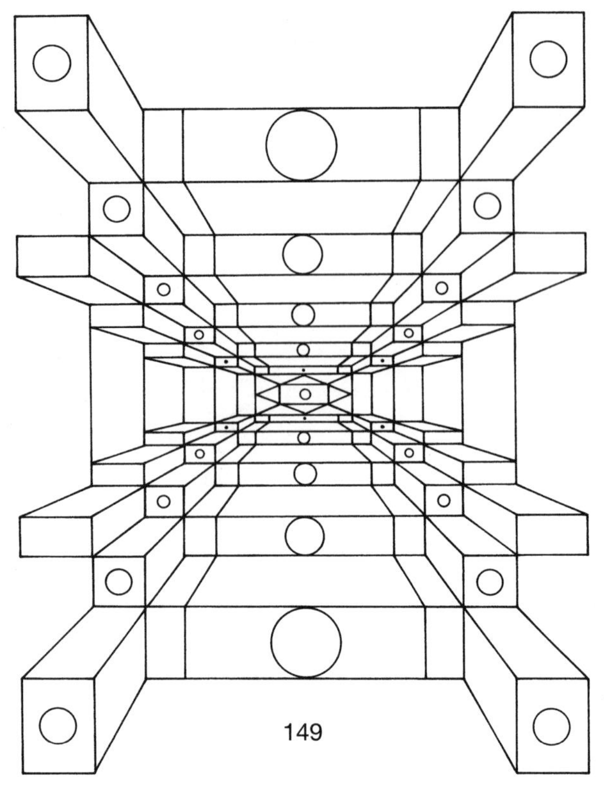

149

150

151

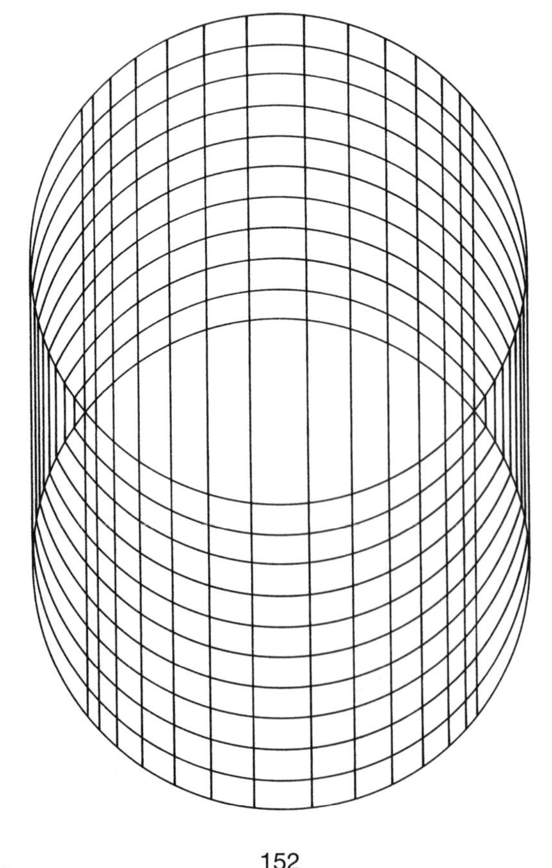

152

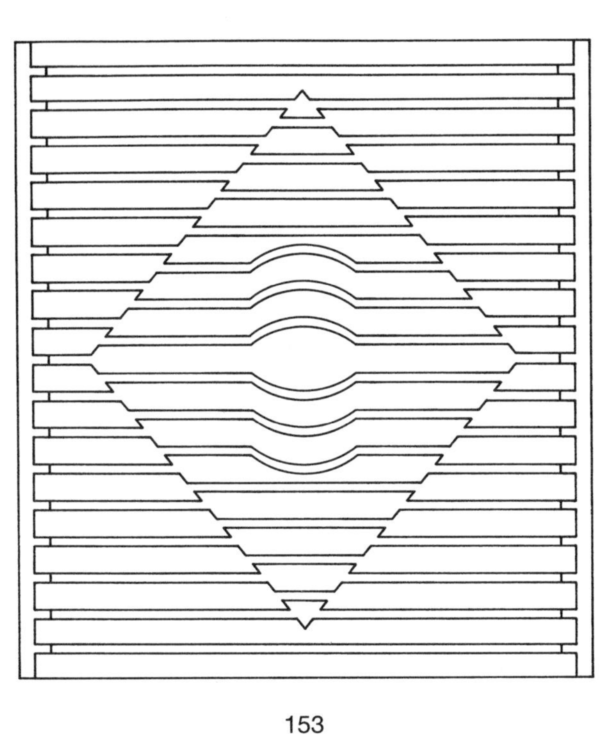

153

154

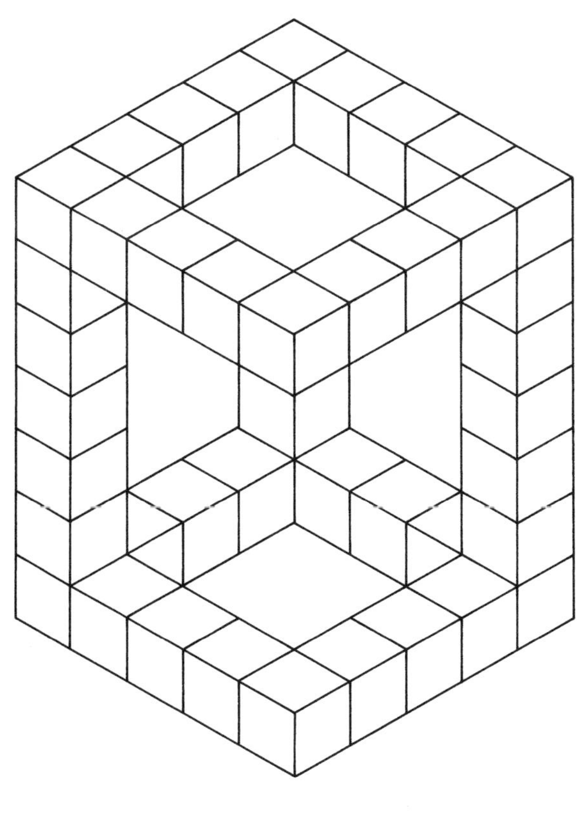

155

156

157

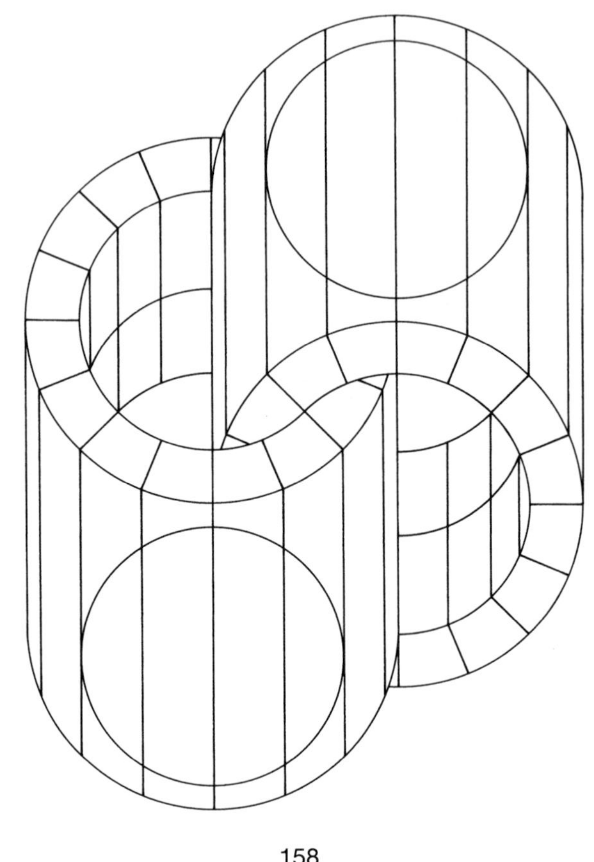

158

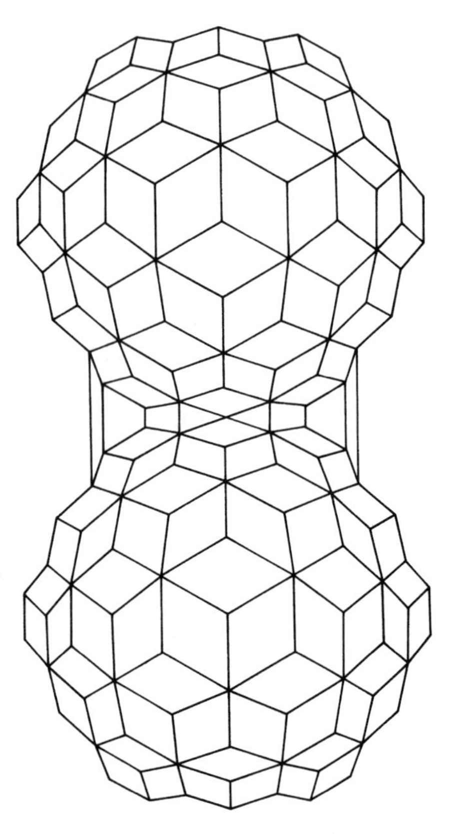

159

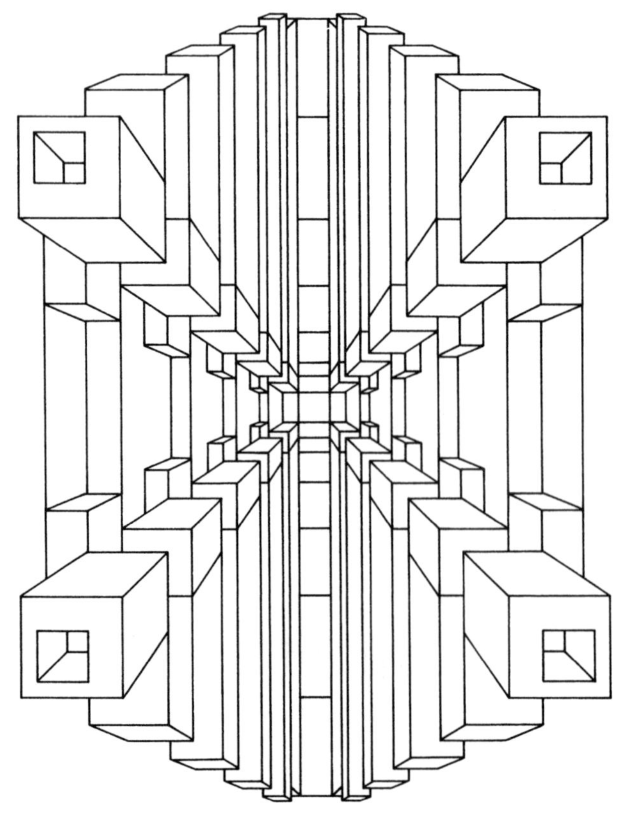

160

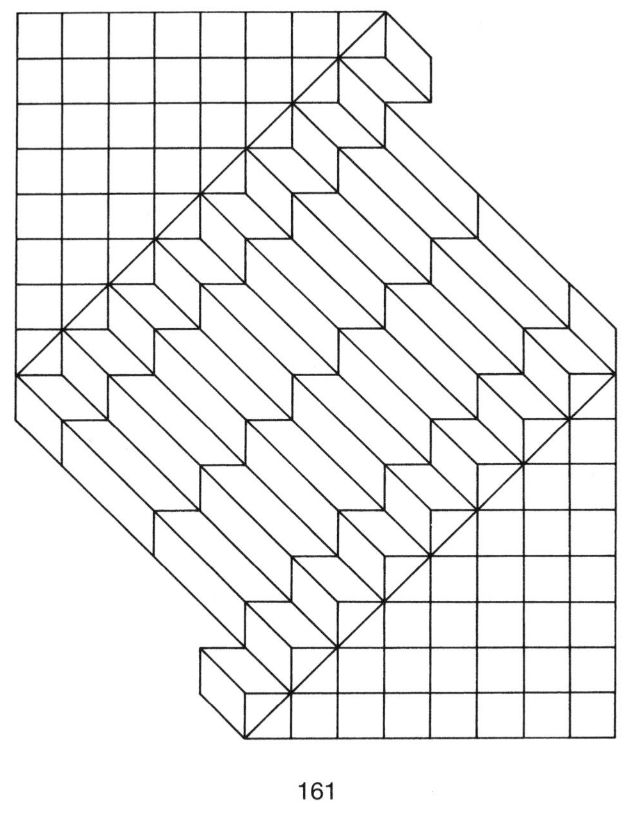

161

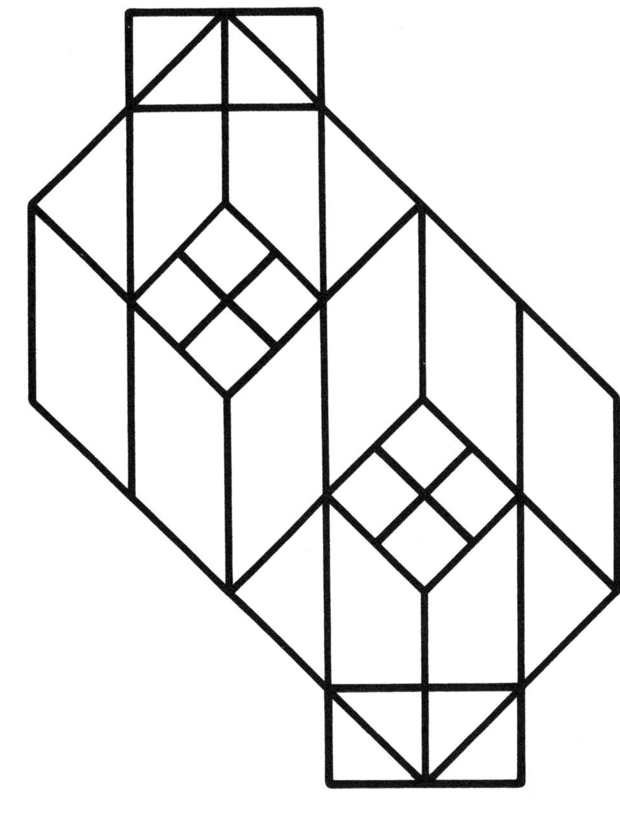

162

163

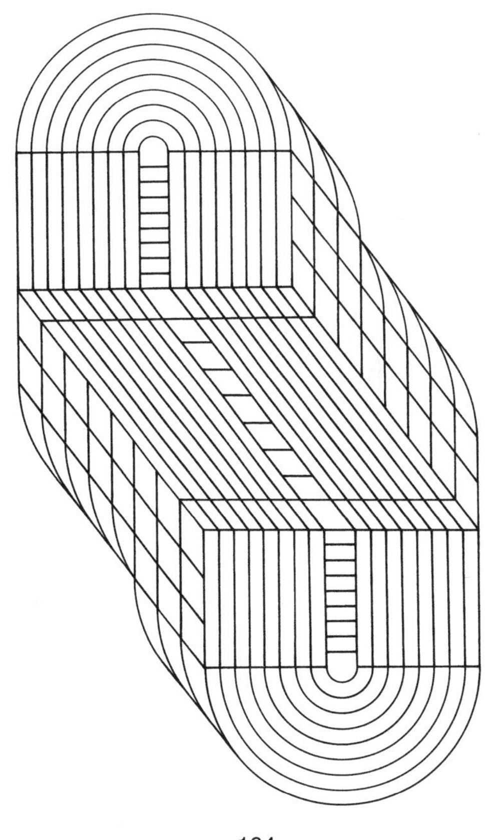

164

165

166

167

168

169

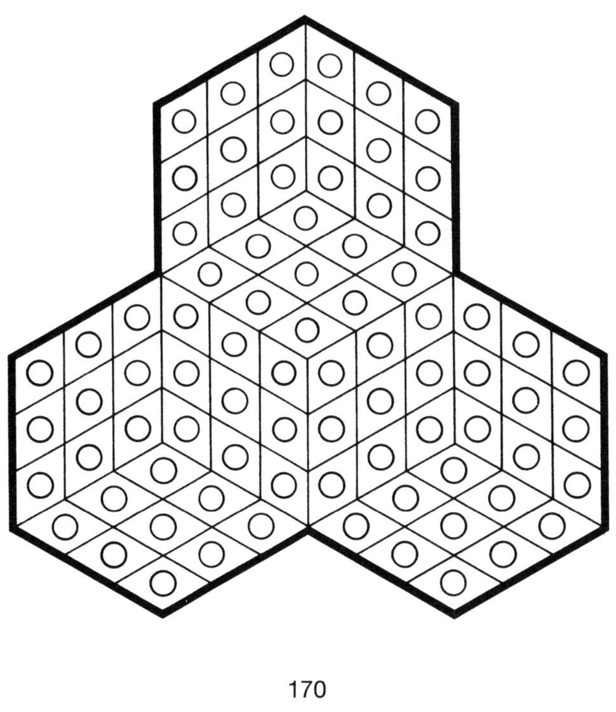

170

171

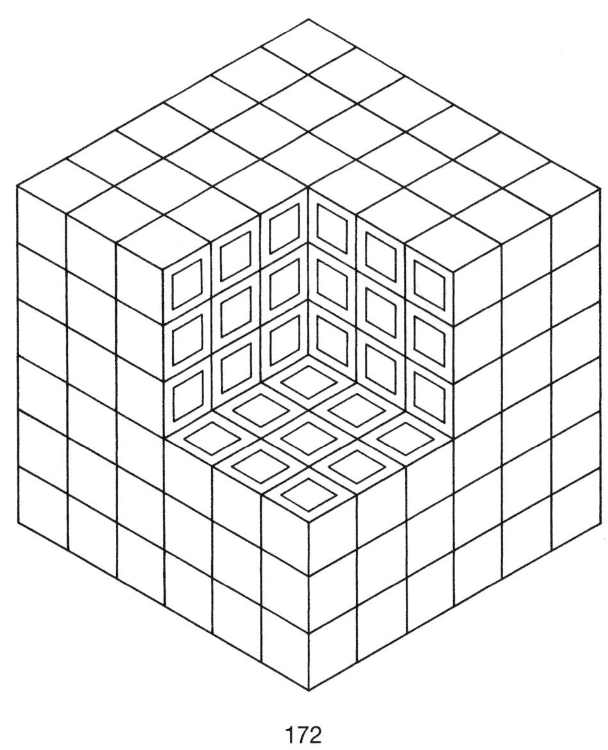

172

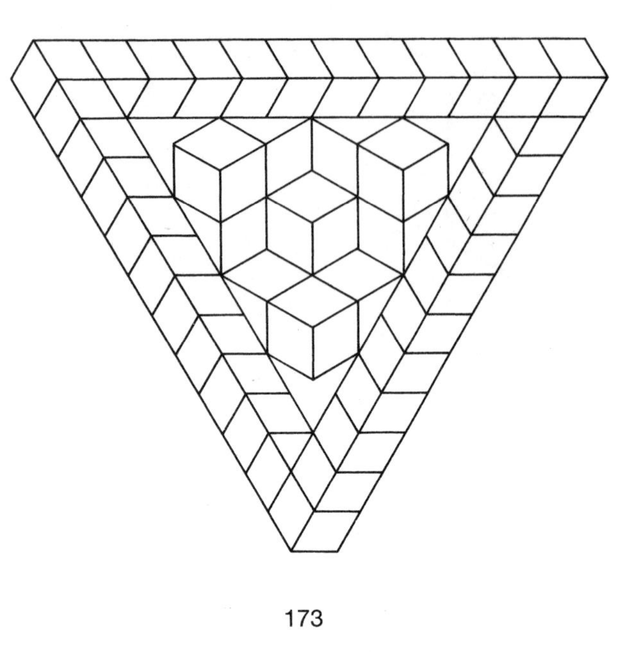

173

174

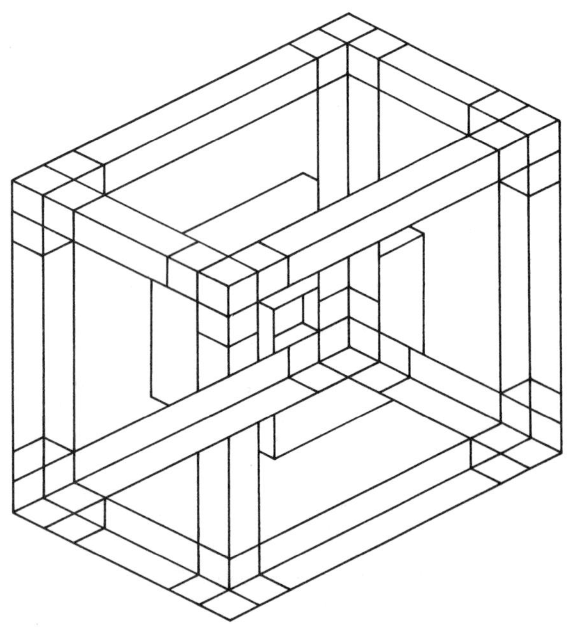

175

176